IS THE MOON LANDING CONSPIRACY TRUE?

J.W. Adams

J.W. Adams

To Teresa

We both have our own fascination with the Moon.

Education and Patriotism

We Americans are taught in school, "The United States went to the Moon a long time ago, but it was so barren, totally nothing to see on the Moon, and way too expensive to go back." I have used that logic before. It seems as if we are conditioned to say that.

How could the Moon, over the last several decades, have been the only field of science where we stopped going further? We regressed in our exploration of the Moon, and our excuses were:

1. It is too dangerous to keep going back.

2. There is really no good reason to return; it is all the same thing on the surface of the Moon.

3. It is just way too expensive.

If these are the reasons no other country has ever gone to the Moon in almost 50 years, I would like to present the following:

Is the Moon Landing Conspiracy True?

1. If it is so dangerous, how did *no one ever die* going to the Moon? They refer to the deaths of the men in the practice flight. That doesn't prove that *getting there* is dangerous.

In fact, it didn't look *dangerous* at all! They were playing golf on the Moon!

2. Too boring up there? Just a bunch of Moon rocks? Nothing to see there? Are you kidding me?

You went through a lot of trouble to get there, landed on a tiny spot where you played golf, did very little science because you were not scientists, you were military patriots, and you determined it is just too pointless to go back?

Yup, no reason to go back, we saw everything it has to offer.

3. Too expensive? Really? Billionaires are now spending whatever it takes to get into low-earth orbit, but it is far too costly to replicate what NASA did with 1960's technology in an aluminum-riveted lander? It seems like it couldn't be *that* difficult to get the opportunity to

play golf on the Moon. What billionaire explorer would not have tried that by now?

Or, hear me out, maybe they *never went* because it *is* way too dangerous and expensive. If I'm right, it is one of the most successful cover-ups of all time. Even the most respected scientists in the country seem to be playing along.

If it were a massive hoax, it would mean President Nixon was responsible for a lot more than Watergate. The U.S. had enough motive to assert global dominance during the Cold War—even the ultimate propaganda film. What better way to reinforce its image of superiority than to send the world's most powerful rockets into outer space, which, if you remember, were always broadcast live during the 1980's.

I remember NASA launches as a kid because they often preempted regular television programming. But how real was it? The motives for staging the Moon landings were perhaps defensible, but I am blown away by reasonable people's unwillingness to even discuss it. It always seems that I conclude, "Eh, what does it matter anyway?" I have personally decided that it does matter. The truth matters.

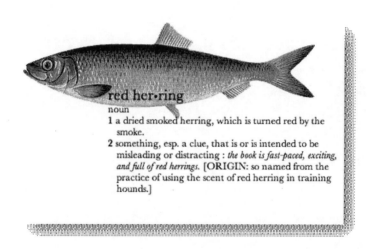

I am not saying the rockets were faked. They were real rockets. There were millions of eye-witnesses to each launch. I do *not* suggest they faked the launching of giant rockets. That is a red herring.

The flag-waving on the Moon controversy—a red herring. Shadow lines are a red herring. No stars in the Moon's sky—red herring. These have all been publically debunked for years, to some extent, so there is plenty of material the patriotic public can watch to help keep itself fooled.

There are countless forms of disinformation out there. "People who believe it was faked say the flag was waving. It wasn't waving; it was bouncing on an aluminum rod. Therefore, your entire argument is disregarded."

No legitimate Moon landing conspiracy theorist (what a beautiful oxymoron) would use the flag waving as evidence. It would be childish to base your whole conspiracy theory on a "flag waving" that probably *was* just bouncing.

There is a ton of legitimate evidence on YouTube these days, mixed in with the obviously-ridiculous reasons that are easy to shrug off.

Is the Moon Landing Conspiracy True?

Think of the flat-earth conspiracy. I wouldn't be surprised if NASA were actively lumping Moon-landing conspiracy videos with flat-earth videos on YouTube. They may have even cooked-up the whole flat-earth idea. If you can get through the swamp of disinformation online, I promise you a very rewarding conspiracy theory journey. We did fake it. I took the blue pill.

https://www.youtube.com/watch?v=J08JkD8NFxo

https://www.youtube.com/watch?v=eicNiSilWUs

J.W. Adams

Is the Moon Landing Conspiracy True?

"In a world of universal deceipt,
telling the truth is a revolutionary act."

George Orwell
1984

First Attempts

This is the third incarnation of the same book. I used the pen name *Teddy Freeman* for the first two. I was convinced that I was onto something so potentially-sensitive that I might face censorship. I know the United States has seemingly always been the champion of free speech, and if I were right, it might carry consequences.

As a frequent teacher of American history, I am well aware of our nation's potential for deception and the corresponding necessity to retain control of the conversation when necessary. In the most extreme cases, I would imagine that they have had to appeal to the scientist's sense of patriotism.

If we did *not* go to the Moon, it is disappointing for all of humanity. The political consequences would be serious—the scientific repercussions even more so. Universal acceptance of this fact will one day make us question everything we have ever seen on television. It also might make the world audience doubt every celebrity-level scientist in the future. My respect for Bill Nye and Neil deGrasse Tyson has been significantly shaken, not because they believe the Moon landing was real but because their reasoning seems deliberately simple.

According to Moon landing believers, the footage was impossible to fake. They present this as evidence of the Apollo missions being genuine. Skill for analyzing any evidence that lies outside of conventional wisdom is curiously missing when famous television scientists are asked about the Apollo missions. This must be what happens when you spend years defending every detail of "amazing science" and very little time exercising critical analysis.

These guys are not the right people to ask anyway. In my opinion, they have as much credibility on this subject as the guys on *Mythbusters*.

I have been asked many times, "What does it matter anyway?" Think about the gravity of that question. Science would be affected for years to come.

Scientists believe the Moon once had a magnetic field, based on the alignment of the metals in the Moon rocks brought back by Apollo. Isn't this much easier to explain once we realize that the Moon rocks collected by NASA are *Earth* rocks?

Is the Moon Landing Conspiracy True?

"In the 1970s, scientists subjected newly arrived samples of Moon rocks to a barrage of tests. To their surprise, they discovered that some of the rocks were magnetic. When magma cooled and solidified into these rocks, the material had been exposed to a magnetic field. Which was strange because, as far as scientists knew at the time, the Moon didn't have a global magnetic field like the one that envelops Earth, then or ever. But the rocks suggested otherwise, and many studies of the samples since have shown that the Moon indeed had a magnetic field billions of years ago, perhaps one as strong as Earth's today."

(Marina Koven August 9, 2017)

https://www.theatlantic.com/science/archive/2017/08/apollo-Moon-rocks-heat/536343/

Below is a link to a Bart Sibrel radio interview. If you are seriously interested, as I am, in knowing the truth regarding the Apollo Missions, I recommend that you listen to the entire interview.

https://www.youtube.com/watch?v=4QzdzM469is

Occam's Razor

A principle authored by William of Ockham (1287-1347), a friar, scholastic philosopher, and theologian. In essence, the principle is that when there are two or more competing hypotheses, the conclusion that relies on the fewest assumptions should be chosen.

A Pointless Conspiracy?

I thought the idea of the U.S. government staging the Moon landing was a crazy obsession that was perfect for bored people. In a way, it is. Who really wants to sit around watching hours of NASA footage? Well, I did, and I surprised myself by personally concluding that the United States did, in fact, fake it.

The deciding factor for me was the motive. It was critical for the U.S. to be successful at this endeavor. There was no room for some terrible Moon disaster. As with all space travel, the potential for loss of human life was very real. Curiously, walking on the Moon seemed to be easy for the most powerful country on Earth. You do not have to look very far to see that powerful countries are capable of powerful propaganda.

The United States was desperate to reassert its global power after the embarrassment of Sputnik. In fact, the Russians were showing-up the United States in all sorts of ways: The first human-made satellite, the first man in space, the first man in orbit, the first woman in space, the first dog in space... The list goes on. If the United States tried something as daring as going to the Moon and failed, the repercussions would be enormous.

"It is *possible* to go to the Moon," the scientists of the day had agreed. One little complication, however, would risk the national security of the most influential nation in the world. Part of the United States' advantage since World War II has been our cutting-edge technology.

Is the Moon Landing Conspiracy True?

J.W. Adams

Is the Moon Landing Conspiracy True?

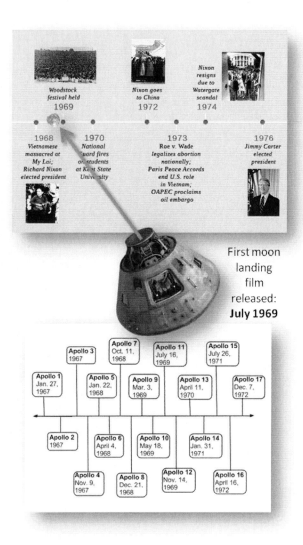

Nixon

Nixon was president at the time. His hands were all over the Apollo Missions. He was likely one of the people who *cooked-up* the whole idea of faking it. If any former president was capable of such deception, it was Richard Nixon.

> http://www.dailymail.co.uk/sciencetech/article-2267391/Remarkable-pictures-NASA-blowing-stretches-Arizona-desert-simulate-surface-Moon.html

Richard Nixon is a no good, lying bastard. He can lie out of both sides of his mouth at the same time, and if he ever caught himself telling the truth, he'd lie just to keep his hand in.

(Harry S. Truman)

Clinton

On page 156 of his autobiography, *My Life*, former president Bill Clinton hints that even he questions if it could have been staged. The language is cleverly disguised; listen to the little breadcrumb he left behind:

"Just a month before, Apollo 11 astronauts Buzz Aldrin and Neil Armstrong had left their colleague, Michael Collins, aboard spaceship Columbia and walked on the Moon…The old carpenter asked me if I really believed it happened. I said sure, I saw it on television. He disagreed; he said that he didn't believe it for a minute, that 'them television fellers' could make things look real that weren't. Back then, I thought he was a crank. During my eight years in Washington, I saw some things on TV that made me wonder if he wasn't ahead of his time." -William J. Clinton, *My Life*

Is the Moon Landing Conspiracy True?

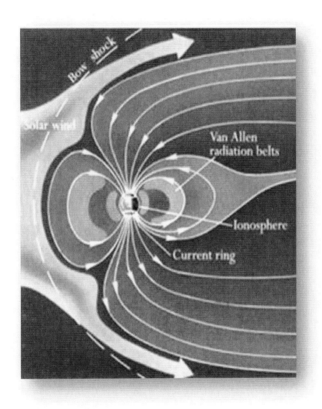

J.W. Adams

Van Allen

The Van Allen radiation belts are two zones of highly-charged particles. They pose a major obstacle to all future space travel, and they are located about 9,300 to 12,400 miles above Earth's surface. NASA itself admits that it would be highly dangerous for a human to pass through that zone.

Not surprisingly, this was never a problem in 1969, or on any of the other Apollo missions. Not only did all of the Apollo astronauts live long and healthy lives, they also never reportedly suffered effects that would be common with radiation poisoning. We might not have had the technology to get through these belts safely in 1969, but we did have the technology to broadcast a simulation.

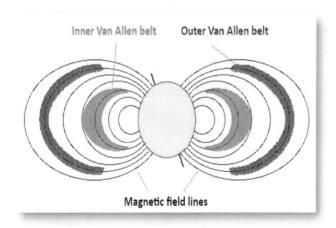

J.W. Adams

Houston

Would a hoax of this magnitude be impossible to pull off? I don't think so at all. The technology was so clunky and ridiculous, how is this not obvious to everyone in the modern age?

Only a few people at the top knew of the deception, including the astronauts. Everyone else: the scientists, the technicians in Houston, and the broadcasters accepting the government's live feed thought it was real, too. So did most of the human race sitting in living rooms and schools across the world. Everyone was watching the same "live feed."

Science or propaganda?

Impossible to fake it?

A Distant Playground

Everyone has seen the video of the men bouncing playfully on the surface of the Moon. They played golf. They drove around in a go-cart that is supposedly still sitting there. They did jumping jacks and all sorts of antics. Excuse me, isn't the Moon lacking an atmosphere, which would mean that even the slightest, tiniest rip in their suit should mean instant death, right? That never comes up as a concern during all the playful stumbling and skipping around on the Moon.

The online video evidence is staggering. The videos analyzed in the links I am providing are official NASA videos. I do not want to go into much more detail because the videos really do speak for themselves.

While searching online, keep in mind that the Moon landing conspiracy videos are often linked to other, usually ridiculous, conspiracy theories. You have to know how to weed out the "flat-earth theory" disinformation—it is everywhere on YouTube. The truth is out there, but stay grounded in logic. The Earth is *not* flat.

Sibrel

I enjoy Bart Sibrel's tenacity. He can be annoying and persistent, but I don't think he's crazy. I think he's right. You can find clips of his movie, *A Funny Thing Happened on the Way to the Moon* on his website, Sibrel.com.

We have to admit that any government, including the one in Washington, D.C., is at least *capable* of pulling off an illusion for the masses.

Is the Moon Landing Conspiracy True?

http://sibrel.com/

https://www.youtube.com/watch?v=xciCJfbTvE4

http://www.thesleuthjournal.com/conspiracy-corner/

Disinformation:

False information deliberately and often covertly spread (as by the planting of rumors) in order to influence public opinion or obscure the truth.

Easily-Overlooked Conveniences

For astronaut training, an exact replica of the sea of tranquility, the landing site of the Apollo missions, was blasted out of the desert in the American southwest. *Isn't that convenient?*

According to NASA, there was a reflector left on the surface of the Moon, regularly used by scientists, making it possible for them to measure the precise distance from the Earth to the Moon. Why doesn't anyone point-out that laser beams can do that with no need for a special mirror on the other end? They measure many near-Earth objects using this method. *Isn't it convenient that the Moon is already an especially-reflective surface.*

Is the Moon Landing Conspiracy True?

Easy to Prove, Either Way

There is one simple way to shut-up all of the conspiracy nuts, like me. Focus one telescope, one time, onto the Sea of Tranquility. Zoom in. Show the visual evidence left behind. That footprint, we have been told, will remain perfectly-in-tact in the vacuum of space. As you might recall, the Moon has no atmosphere, so there is no force of erosion to act upon it.

If we are told that no such telescope exists, I postulate that this is what we would expect to hear if this was a cover-up. Astronomers can zoom-in on details much farther away. The public eye is often kept distracted from the infamous landing spot on the Sea of Tranquility.

Past NASA projects that have traveled near the moon, even though always equipped with the most sophisticated cameras, never thought it was important enough to zoom-in and pay homage to this sacred location of human exploration.

Is the Moon Landing Conspiracy True?

NASA spacecraft has frequently used the Moon to sling-shot into space. Not one single snapshot of that historic site. How odd. One would expect a Moon black-out from a government who is afraid that the public will no longer trust what they see on television.

I know that the *Myth-Busters* guys did a couple of shows on it, but they were chasing the evidence that is easily dismissed. There are so many silly arguments; it is an effective distraction from the real ones, which I have already listed.

J.W. Adams

(c) 2005 IMAX Corporation and Playtone. Photo: Melinda Sue Gordon

Adam Doesn't Ruin Enough

I am a fan of *Adam Ruins Everything*. I feel like Adam Conover has an open mind and is usually seems only concerned with the facts. I had to wonder how logical he was being about the Moon landing when he said, "Given the film-making and lighting technology at the time, [faking the Moon landing] actually wouldn't have been possible."

https://www.youtube.com/watch?v=zhp-FTYSGe8

Not possible? Are you kidding me? Is it so difficult to use common sense on this issue? Again, as if we are all brainwashed, we buy this line. It was *very* possible. To demonstrate, watch the Moon scene from *2001: A Space Odyssey*. To say it was impossible in 1969 is an insult to public intelligence.

Where there is a will to believe, and a public that trusts Richard Nixon, there is a soundstage and some grainy footage that is apparently "impossible" to fake.

https://www.youtube.com/watch?v=oU4Rk0NATNs

Is the Moon Landing Conspiracy True?

Is it so hard to see that it *wasn't impossible* to fake the Moon landing? If you have determined like me, that it **was** possible, then you begin to see that even Adam Conover seems to be plugged into the matrix on this one.

You can only believe that it was impossible if you are willing to set all of your common sense aside in blind faith. A critical analysis will lead to a different conclusion.

I highly suggest that you take no one's word for it. Watch the footage, notice the circular logic used to defend the Apollo missions, and then decide for yourself.

https://www.youtube.com/watch?v=AsCqsJDpHHU

Justifiable?

Even if this was the greatest televised illusion ever to fool the world, it was as brilliant as it was repulsive. If the United States government *did* fake the Moon-landing footage from 1969-1972, then it was a genius strategy, albeit an immoral and ethically-reprehensible one.

When the propaganda film is viewed in conjunction with the entire U.S. space program, which is tightly linked with the U.S. military, one could argue that 1969 marked the beginning of the end of the Cold War.

The Apollo Mission era was the foundation upon which the United States fashioned its forward-leaning persona of the late 20th century. I do not wish to get into the controversies put forth online about NASA continuing to embezzle billions by fooling the viewing public, but there is some interesting evidence on YouTube and elsewhere.

Is the Moon Landing Conspiracy True?

I am a little suspicious of everything on YouTube, as everyone should be. Modern NASA trickery videos could be another red herring being encouraged by the Powers That Be to steer attention away from the deception of the Apollo Missions.

The U.S.S.R. struggled to overcome the juggernaut momentum of the United States in the decades following Armstrong's famous words. Eventually, the Cold War ground to a halt.

After the fall of the U.S.S.R., the United States could have used the opportunity to explain the necessity of the deception, but perhaps such an admission would have seemed like a weakness. Would we ever believe such an event again?

If the government were to admit to such deception, would we have spent the end of the 20th century distrusting everything we see on television? The 9/11 conspiracy theories are evidence that many people are already there. For the record, I do not believe anything other than the generally-accepted events of that tragedy.

The difference between 9/11 and the Moon landing conspiracies is the presence of a clear motive. I have heard the motives presented by 9/11 truthers. I do not find them compelling. They are too complex, and the death toll was in the thousands. Saying it was a deliberate act by anyone other than terrorists is too much of a stretch for me.

The Apollo missions were different. One could argue that faking the Moon landing saved lives by ending the Cold War. This would make it justifiable to a lot of people.

The Patriots

I believe that Buzz Aldrin, Neil Armstrong, and the other patriotic Americans involved, were good people. Their motives were good. In my opinion, they probably had good reason to feel that the deception was justified. Why, after all these years, has it not occurred to any of them that it might be time to tell the truth?

I believe that Armstrong was possibly told that the truth would come out eventually. When it didn't, he preferred to stay off the topic. Viewed as a "reluctant American hero," Armstrong stayed out of the public eye for much of his life. Moon conspiracy theorists believe that his conscience would not let him continue the deception.

"To you we say we have only completed a beginning. We leave you much that is undone.

Is the Moon Landing Conspiracy True?

There are great ideas undiscovered, breakthroughs available to those who can remove one of the truth's protective layers."
-Neil Armstrong, 1994

The life of a recluse is odd behavior for an explorer who was brave enough to travel 238,000 miles away and become the first man to walk on the Moon. Would a man of his character not have felt an obligation to continue answering questions until his dying day? Men who have been to the Moon are not exactly commonplace.

His descriptions should have been invaluable—as with all of the astronauts who participated in the other Apollo missions. Interestingly, the astronauts' stories all include a very clear opinion that it was a barren landscape with nothing left to learn from.

Search for the Truth

You do not have to take my word for it. As with any conspiracy theory, I suggest you examine all of the evidence before drawing a conclusion. I believe mine has been a very open-minded approach. If you are determined to prove it one way or another, you must set your prejudices aside and watch all of the evidence being presented.

When I Googled, "Proof we did land on the Moon," I systematically examined every piece of evidence I could find. It was not easy to find very much. You would expect irrefutable evidence such as a satellite image of the landing site or an undisputed Moon rock.

In my first search result, I see that The Weather Channel has some proof. I was hoping to see some satellite imagery on a site like that. Instead, it is a story of a director who claims that the images seen in 1969 could not have been staged. Really? This is such a ridiculous argument. Use your most basic common sense as an intelligent adult. Did they not possess the technology to *stage* the Moon landing in 1969? Of course, they did! One of the biggest movies of 1968 was *2001: A Space Odyssey*, which depicted the moonscape and movement in low gravity very accurately.

None of the NASA Moon footage is of excellent picture quality. Such a momentous occasion, you think they would have had the best cameras in the world.

https://weather.com/news/news/Moon-landing-proven-20130123

The *Star Trek* pilot aired in 1965

The excuses for why they only produced grainy footage are laughable. Again, you will have to expect some very poor explanation about how heavy cameras were back then, or how that was the best technology at the time.

If better cameras were so heavy, you really should have left that lunar rover behind. It served no purpose other than as a prop for Moon stunts. The entire world would have benefited from better Moon *footage*. It was inspirational to millions.

The film would have been examined in detail for generations to come, but NASA has "lost" all film that has not already been publically released.

Popular Science was the next site in the search results. I had very high hopes for some concrete evidence. Sadly, even Popular Science only offered another "analysis" of the footage. It was not even compelling. Such camera tricks were common even in 1969, so please give me a shred of irrefutable evidence.

There must be air-tight proof *somewhere*! I mean, the United States was doing it to *prove* we could be the first to do it! You would expect the astronauts to bring back more than just photographs and rocks that look exactly like broken meteorites. *Oh, that's right, there was nothing there but rocks.*

Popular Science provides a painfully-unhelpful article that gets into the technicalities of how the Moon dust kicked up by the rover could only have happened on the Moon. Not anywhere near the proof I am looking for.

https://www.popsci.com/blog-network/vintage-space/proof-we-landed-Moon-dust#page-3

In the next search result, some website called "Universe Today" hurls insults at conspiracy theorists while offering no good proof either. Actually, they offer no proof at all. They point to the fact that thousands and thousands of people worked on the Apollo projects. As I have pointed out many times, everyone was working on what they *thought* was a legitimate project. Everyone in that control room *believed* it was real.

The science was real *enough*, but it is still possible that the rocket carried no human beings and the transmission being received, including by NASA's own control room, was being directed

Is the Moon Landing Conspiracy True?

by a select few from a soundstage. There was very little difference between the simulations and the actual "event."

https://www.universetoday.com/111188/how-do-we-know-the-Moon-landing-isnt-fake/

Every online result I found was the same. You can tell that the public wants so badly for it to be true. The evidence is clearly as refutable as ever, but the believers far outnumber the skeptics.

Notice how unscientific the most popular arguments are. NASA has left everyone out there dangling, desperate to use the same footage to prove that it *did* happen. NASA still says they definitely walked on the Moon during the Apollo era. Why is there no other proof besides grainy video and high-definition snapshots of astronauts posing? There is no other proof, and it is as easy to explain away as a Bigfoot sighting.

I offer a warning about trying to find proof online. It is painful. NASA offers *nothing* irrefutable. The videos made public by NASA give us significant reason to doubt. We know it was not impossible to fake; we are not stupid.

"If you were to stand on a street corner and ask each passerby to tell you what a fact is, most people would tell you that a fact is what is real and true. However, this common notion is mistaken. Facts are our interpretation of what is real and true.

Yet the problem with interpretations is that they can be dead wrong. Human history provides us with many examples... Human beings need facts because they need certainties in order to proceed through the world. But we should not forget that human beings are fallible."

Marlys Mayfield
Thinking for Yourself, 2007
College of Alameda
Thompson-Wadsworth

NASA's Proof

What evidence does the government use to quiet the critics? I will list their arguments below. Examine for yourself if this evidence is irrefutable. According to the United States government, we should believe that man walked on the Moon in 1969 because:

1. Everyone watched it live.

2. It would be too difficult to keep such a secret for so long in a free country.

3. They brought back Moon rocks.

4. Your grandpa remembers it.

5. You can believe what you want to believe.

6. What does it matter anyway?

Does the truth matter to **you**?

Is the Moon Landing Conspiracy True?

J.W. Adams

Is the Moon Landing Conspiracy True?

Made in United States
North Haven, CT
13 February 2025

65659463R00052